# My Bucket List Journal

# Why A Bucket List Journal?

First let's define what a bucket list is. It is a list of 100 things you want to do before you die or kick the bucket as they say. It can be things like traveling, meeting someone famous or doing something that scares the you-know-what out of you.

Instead of boring you with studies I will just let you know why I think you need to have a bucket list.

It allows you to focus on why you want to do something and are not so distracted on the how. We are so busy trying to figure out how to make more money, spend time with loved ones or find happiness that we forget why we are doing it all in the first place. If you forget your why your goals lose their meaning. The 'why' defines your purpose and fuels your motivation to persevere regardless of the challenges along the way.

By writing out your goals on a bucket list, you can actually take action on them. You see right in front of you that it is your dream to attend an NFL game so you start hunting for tickets. Maybe it is to travel Europe for six months but not for five more years. You can still start reading up on Europe so by the time you get there you know exactly what you want to do and why. Researchers find that anticipating an experience gives us almost as much of a thrill as the actual event.

The most important reason for having a bucket list is to ultimately achieve your goals. To live a meaningful life doesn't mean that you did everything you set out to achieve but that you tried. Instead of wondering what could have been, you will have created some great memories because you actually did something with your life. Something just for you or in this case 100 things just for you.

These memories are the basis of a life that is meaningful, not for others, but for yourself. Acquiring memories is for more fulfilling than chasing material possessions. If you want to look at a bucket list in a different way it could be that you want to create 100 memories you can share before you die.

Dream big and live even bigger.

Bucket List Goal #1: ______________________

Describe The Goal:

______________________

______________________

______________________

______________________

______________________

Why I Want To Accomplish This Goal:

______________________

______________________

______________________

______________________

______________________

By When: ______________________

What do I have to do to achieve this goal?

______________________

______________________

______________________

______________________

Date Achieved: ____________________

What I Felt When I Accomplished My Goal:

____________________

____________________

____________________

____________________

____________________

____________________

____________________

Paste Your Achievement Photo Here

Bucket List Goal #2: ______________________

Describe The Goal:

______________________

______________________

______________________

______________________

______________________

Why I Want To Accomplish This Goal:

______________________

______________________

______________________

______________________

______________________

By When: ______________________

What do I have to do to achieve this goal?

______________________

______________________

______________________

______________________

Date Achieved: ______________________________

What I Felt When I Accomplished My Goal:

______________________________________________

______________________________________________

______________________________________________

______________________________________________

______________________________________________

______________________________________________

______________________________________________

Paste Your Achievement Photo Here

Bucket List Goal #3: ______________________

Describe The Goal:

______________________

______________________

______________________

______________________

______________________

Why I Want To Accomplish This Goal:

______________________

______________________

______________________

______________________

______________________

By When: ______________________

What do I have to do to achieve this goal?

______________________

______________________

______________________

______________________

Date Achieved: ______________________

What I Felt When I Accomplished My Goal:

______________________

______________________

______________________

______________________

______________________

______________________

______________________

Paste Your Achievement Photo Here

Bucket List Goal #4: ______________________________

Describe The Goal:

______________________________________________________

______________________________________________________

______________________________________________________

______________________________________________________

______________________________________________________

Why I Want To Accomplish This Goal:

______________________________________________________

______________________________________________________

______________________________________________________

______________________________________________________

______________________________________________________

By When: ________________________________________

What do I have to do to achieve this goal?

______________________________________________________

______________________________________________________

______________________________________________________

______________________________________________________

Date Achieved: ______________________________

What I Felt When I Accomplished My Goal:

___________________________________________

___________________________________________

___________________________________________

___________________________________________

___________________________________________

___________________________________________

___________________________________________

Paste Your Achievement Photo Here

Bucket List Goal #5: ________________________

Describe The Goal:

________________________________________

________________________________________

________________________________________

________________________________________

________________________________________

Why I Want To Accomplish This Goal:

________________________________________

________________________________________

________________________________________

________________________________________

________________________________________

By When: ____________________________

What do I have to do to achieve this goal?

________________________________________

________________________________________

________________________________________

________________________________________

Date Achieved: ______________________________

What I Felt When I Accomplished My Goal:

____________________________________________

____________________________________________

____________________________________________

____________________________________________

____________________________________________

____________________________________________

____________________________________________

Paste Your Achievement Photo Here

Bucket List Goal #6: ______________________

Describe The Goal:

______________________________________

______________________________________

______________________________________

______________________________________

______________________________________

Why I Want To Accomplish This Goal:

______________________________________

______________________________________

______________________________________

______________________________________

______________________________________

By When: ______________________________

What do I have to do to achieve this goal?

______________________________________

______________________________________

______________________________________

______________________________________

Date Achieved: ____________________

What I Felt When I Accomplished My Goal:

____________________

____________________

____________________

____________________

____________________

____________________

____________________

Paste Your Achievement Photo Here

Bucket List Goal #7: ______________________

Describe The Goal:

______________________________________

______________________________________

______________________________________

______________________________________

______________________________________

Why I Want To Accomplish This Goal:

______________________________________

______________________________________

______________________________________

______________________________________

______________________________________

By When: ______________________________

What do I have to do to achieve this goal?

______________________________________

______________________________________

______________________________________

______________________________________

Date Achieved: ________________________________

What I Felt When I Accomplished My Goal:

________________________________________________

________________________________________________

________________________________________________

________________________________________________

________________________________________________

________________________________________________

________________________________________________

Paste Your Achievement Photo Here

Bucket List Goal #8: ______________________

Describe The Goal:

______________________________________

______________________________________

______________________________________

______________________________________

______________________________________

Why I Want To Accomplish This Goal:

______________________________________

______________________________________

______________________________________

______________________________________

______________________________________

By When: ______________________________

What do I have to do to achieve this goal?

______________________________________

______________________________________

______________________________________

______________________________________

Date Achieved: ____________________

What I Felt When I Accomplished My Goal:

____________________

____________________

____________________

____________________

____________________

____________________

____________________

Paste Your Achievement Photo Here

Bucket List Goal #9: ______________________

Describe The Goal:

______________________________________

______________________________________

______________________________________

______________________________________

______________________________________

Why I Want To Accomplish This Goal:

______________________________________

______________________________________

______________________________________

______________________________________

______________________________________

By When: ______________________________

What do I have to do to achieve this goal?

______________________________________

______________________________________

______________________________________

______________________________________

Date Achieved: ______________________________

What I Felt When I Accomplished My Goal:

____________________________________________

____________________________________________

____________________________________________

____________________________________________

____________________________________________

____________________________________________

____________________________________________

Paste Your Achievement Photo Here

Bucket List Goal #10: ______________________

Describe The Goal:

________________________________________

________________________________________

________________________________________

________________________________________

________________________________________

Why I Want To Accomplish This Goal:

________________________________________

________________________________________

________________________________________

________________________________________

________________________________________

By When: ______________________________

What do I have to do to achieve this goal?

________________________________________

________________________________________

________________________________________

________________________________________

Date Achieved: ______________________

What I Felt When I Accomplished My Goal:

______________________

______________________

______________________

______________________

______________________

______________________

______________________

Paste Your Achievement Photo Here

Bucket List Goal #11: ____________________

Describe The Goal:

______________________________

______________________________

______________________________

______________________________

______________________________

Why I Want To Accomplish This Goal:

______________________________

______________________________

______________________________

______________________________

______________________________

By When: ____________________

What do I have to do to achieve this goal?

______________________________

______________________________

______________________________

______________________________

Date Achieved: ______________________________

What I Felt When I Accomplished My Goal:

______________________________________________

______________________________________________

______________________________________________

______________________________________________

______________________________________________

______________________________________________

______________________________________________

Paste Your Achievement Photo Here

Bucket List Goal #12: ______________________

Describe The Goal:

______________________________________

______________________________________

______________________________________

______________________________________

______________________________________

Why I Want To Accomplish This Goal:

______________________________________

______________________________________

______________________________________

______________________________________

______________________________________

By When: ______________________________

What do I have to do to achieve this goal?

______________________________________

______________________________________

______________________________________

______________________________________

Date Achieved: ______________________________

What I Felt When I Accomplished My Goal:

______________________________________________

______________________________________________

______________________________________________

______________________________________________

______________________________________________

______________________________________________

______________________________________________

Paste Your Achievement Photo Here

Bucket List Goal #13: ____________________

Describe The Goal:

______________________________

______________________________

______________________________

______________________________

______________________________

Why I Want To Accomplish This Goal:

______________________________

______________________________

______________________________

______________________________

______________________________

By When: ____________________

What do I have to do to achieve this goal?

______________________________

______________________________

______________________________

______________________________

Date Achieved: ______________________________

What I Felt When I Accomplished My Goal:

______________________________

______________________________

______________________________

______________________________

______________________________

______________________________

______________________________

Paste Your Achievement Photo Here

Bucket List Goal #14: ____________________

Describe The Goal:

______________________________

______________________________

______________________________

______________________________

______________________________

Why I Want To Accomplish This Goal:

______________________________

______________________________

______________________________

______________________________

______________________________

By When: ____________________

What do I have to do to achieve this goal?

______________________________

______________________________

______________________________

______________________________

Date Achieved: ______________________

What I Felt When I Accomplished My Goal:

______________________

______________________

______________________

______________________

______________________

______________________

______________________

Paste Your Achievement Photo Here

Bucket List Goal #15: ______________________

Describe The Goal:

______________________________________

______________________________________

______________________________________

______________________________________

______________________________________

Why I Want To Accomplish This Goal:

______________________________________

______________________________________

______________________________________

______________________________________

______________________________________

By When: ______________________________

What do I have to do to achieve this goal?

______________________________________

______________________________________

______________________________________

______________________________________

Date Achieved: ______________________________

What I Felt When I Accomplished My Goal:

______________________________________________

______________________________________________

______________________________________________

______________________________________________

______________________________________________

______________________________________________

______________________________________________

Paste Your Achievement Photo Here

Bucket List Goal #16: ______________________

Describe The Goal:

______________________________________

______________________________________

______________________________________

______________________________________

______________________________________

Why I Want To Accomplish This Goal:

______________________________________

______________________________________

______________________________________

______________________________________

______________________________________

By When: ______________________________

What do I have to do to achieve this goal?

______________________________________

______________________________________

______________________________________

______________________________________

Date Achieved: ______________________

What I Felt When I Accomplished My Goal:

______________________

______________________

______________________

______________________

______________________

______________________

______________________

Paste Your Achievement Photo Here

Bucket List Goal #17: ______________________

Describe The Goal:

________________________________________

________________________________________

________________________________________

________________________________________

________________________________________

Why I Want To Accomplish This Goal:

________________________________________

________________________________________

________________________________________

________________________________________

________________________________________

By When: ______________________________

What do I have to do to achieve this goal?

________________________________________

________________________________________

________________________________________

________________________________________

Date Achieved: ______________________________

What I Felt When I Accomplished My Goal:

______________________________________________

______________________________________________

______________________________________________

______________________________________________

______________________________________________

______________________________________________

______________________________________________

Paste Your Achievement Photo Here

Bucket List Goal #18: ______________________

Describe The Goal:

______________________________________________

______________________________________________

______________________________________________

______________________________________________

______________________________________________

Why I Want To Accomplish This Goal:

______________________________________________

______________________________________________

______________________________________________

______________________________________________

______________________________________________

By When: ______________________________

What do I have to do to achieve this goal?

______________________________________________

______________________________________________

______________________________________________

______________________________________________

Date Achieved: ______________________________

What I Felt When I Accomplished My Goal:

_____________________________________________

_____________________________________________

_____________________________________________

_____________________________________________

_____________________________________________

_____________________________________________

_____________________________________________

Paste Your Achievement Photo Here

Bucket List Goal #19: ______________________

Describe The Goal:

________________________________________

________________________________________

________________________________________

________________________________________

________________________________________

Why I Want To Accomplish This Goal:

________________________________________

________________________________________

________________________________________

________________________________________

________________________________________

By When: ______________________________

What do I have to do to achieve this goal?

________________________________________

________________________________________

________________________________________

________________________________________

Date Achieved: ______________________________

What I Felt When I Accomplished My Goal:

______________________________

______________________________

______________________________

______________________________

______________________________

______________________________

______________________________

Paste Your Achievement Photo Here

Bucket List Goal #20: ______________________

Describe The Goal:

______________________________________

______________________________________

______________________________________

______________________________________

______________________________________

Why I Want To Accomplish This Goal:

______________________________________

______________________________________

______________________________________

______________________________________

______________________________________

By When: ______________________________

What do I have to do to achieve this goal?

______________________________________

______________________________________

______________________________________

______________________________________

Date Achieved: ______________________

What I Felt When I Accomplished My Goal:

______________________________________________

______________________________________________

______________________________________________

______________________________________________

______________________________________________

______________________________________________

______________________________________________

Paste Your Achievement Photo Here

Bucket List Goal #21: ______________________

Describe The Goal:

________________________________________

________________________________________

________________________________________

________________________________________

________________________________________

Why I Want To Accomplish This Goal:

________________________________________

________________________________________

________________________________________

________________________________________

________________________________________

By When: ______________________________

What do I have to do to achieve this goal?

________________________________________

________________________________________

________________________________________

________________________________________

Date Achieved: ______________________________

What I Felt When I Accomplished My Goal:

______________________________________________

______________________________________________

______________________________________________

______________________________________________

______________________________________________

______________________________________________

______________________________________________

Paste Your Achievement Photo Here

Bucket List Goal #22: ______________________

Describe The Goal:

________________________________________

________________________________________

________________________________________

________________________________________

________________________________________

Why I Want To Accomplish This Goal:

________________________________________

________________________________________

________________________________________

________________________________________

________________________________________

By When: ______________________________

What do I have to do to achieve this goal?

________________________________________

________________________________________

________________________________________

________________________________________

Date Achieved: ______________________________

What I Felt When I Accomplished My Goal:

_____________________________________________

_____________________________________________

_____________________________________________

_____________________________________________

_____________________________________________

_____________________________________________

_____________________________________________

Paste Your Achievement Photo Here

Bucket List Goal #23: ______________________

Describe The Goal:

______________________________________

______________________________________

______________________________________

______________________________________

______________________________________

Why I Want To Accomplish This Goal:

______________________________________

______________________________________

______________________________________

______________________________________

______________________________________

By When: ______________________

What do I have to do to achieve this goal?

______________________________________

______________________________________

______________________________________

______________________________________

Date Achieved: ______________________

What I Felt When I Accomplished My Goal:

______________________

______________________

______________________

______________________

______________________

______________________

______________________

Paste Your Achievement Photo Here

Bucket List Goal #24: ______________________

Describe The Goal:

______________________________________

______________________________________

______________________________________

______________________________________

______________________________________

Why I Want To Accomplish This Goal:

______________________________________

______________________________________

______________________________________

______________________________________

______________________________________

By When: ______________________________

What do I have to do to achieve this goal?

______________________________________

______________________________________

______________________________________

______________________________________

Date Achieved: ____________________

What I Felt When I Accomplished My Goal:

____________________

____________________

____________________

____________________

____________________

____________________

____________________

Paste Your Achievement Photo Here

Bucket List Goal #25: ______________________

Describe The Goal:

______________________________________

______________________________________

______________________________________

______________________________________

______________________________________

Why I Want To Accomplish This Goal:

______________________________________

______________________________________

______________________________________

______________________________________

______________________________________

By When: ______________________________

What do I have to do to achieve this goal?

______________________________________

______________________________________

______________________________________

______________________________________

Date Achieved: ______________________________

What I Felt When I Accomplished My Goal:

______________________________

______________________________

______________________________

______________________________

______________________________

______________________________

______________________________

Paste Your Achievement Photo Here

Bucket List Goal #26: ______________________

Describe The Goal:

______________________________________

______________________________________

______________________________________

______________________________________

______________________________________

Why I Want To Accomplish This Goal:

______________________________________

______________________________________

______________________________________

______________________________________

______________________________________

By When: ______________________________

What do I have to do to achieve this goal?

______________________________________

______________________________________

______________________________________

______________________________________

Date Achieved: ______________________________

What I Felt When I Accomplished My Goal:

______________________________________________

______________________________________________

______________________________________________

______________________________________________

______________________________________________

______________________________________________

______________________________________________

Paste Your Achievement Photo Here

Bucket List Goal #27: ______________________

Describe The Goal:

______________________________________

______________________________________

______________________________________

______________________________________

______________________________________

Why I Want To Accomplish This Goal:

______________________________________

______________________________________

______________________________________

______________________________________

______________________________________

By When: ______________________________

What do I have to do to achieve this goal?

______________________________________

______________________________________

______________________________________

______________________________________

Date Achieved: ______________________________

What I Felt When I Accomplished My Goal:

______________________________________________

______________________________________________

______________________________________________

______________________________________________

______________________________________________

______________________________________________

______________________________________________

Paste Your Achievement Photo Here

Bucket List Goal #28: ______________________

Describe The Goal:

______________________

______________________

______________________

______________________

______________________

Why I Want To Accomplish This Goal:

______________________

______________________

______________________

______________________

______________________

By When: ______________________

What do I have to do to achieve this goal?

______________________

______________________

______________________

______________________

Date Achieved: ______________________________

What I Felt When I Accomplished My Goal:

______________________________________________

______________________________________________

______________________________________________

______________________________________________

______________________________________________

______________________________________________

______________________________________________

Paste Your Achievement Photo Here

Bucket List Goal #29: ______________________

Describe The Goal:

________________________________________

________________________________________

________________________________________

________________________________________

________________________________________

Why I Want To Accomplish This Goal:

________________________________________

________________________________________

________________________________________

________________________________________

________________________________________

By When: ______________________________

What do I have to do to achieve this goal?

________________________________________

________________________________________

________________________________________

________________________________________

Date Achieved: ______________________________

What I Felt When I Accomplished My Goal:

______________________________________________

______________________________________________

______________________________________________

______________________________________________

______________________________________________

______________________________________________

______________________________________________

Paste Your Achievement Photo Here

Bucket List Goal #30: ______________________

Describe The Goal:

________________________________________

________________________________________

________________________________________

________________________________________

________________________________________

Why I Want To Accomplish This Goal:

________________________________________

________________________________________

________________________________________

________________________________________

________________________________________

By When: ______________________________

What do I have to do to achieve this goal?

________________________________________

________________________________________

________________________________________

________________________________________

Date Achieved: ___________________________

What I Felt When I Accomplished My Goal:

___________________________________________

___________________________________________

___________________________________________

___________________________________________

___________________________________________

___________________________________________

___________________________________________

Paste Your Achievement Photo Here

Bucket List Goal #31: ______________________

Describe The Goal:

________________________________________

________________________________________

________________________________________

________________________________________

________________________________________

Why I Want To Accomplish This Goal:

________________________________________

________________________________________

________________________________________

________________________________________

________________________________________

By When: ______________________________

What do I have to do to achieve this goal?

________________________________________

________________________________________

________________________________________

________________________________________

Date Achieved: ____________________

What I Felt When I Accomplished My Goal:

____________________

____________________

____________________

____________________

____________________

____________________

____________________

Paste Your Achievement Photo Here

Bucket List Goal #32: ______________________

Describe The Goal:

______________________________________

______________________________________

______________________________________

______________________________________

______________________________________

Why I Want To Accomplish This Goal:

______________________________________

______________________________________

______________________________________

______________________________________

______________________________________

By When: ______________________________

What do I have to do to achieve this goal?

______________________________________

______________________________________

______________________________________

______________________________________

Date Achieved: ______________________________

What I Felt When I Accomplished My Goal:

______________________________

______________________________

______________________________

______________________________

______________________________

______________________________

______________________________

Paste Your Achievement Photo Here

Bucket List Goal #33: ______________________

Describe The Goal:

______________________________________________

______________________________________________

______________________________________________

______________________________________________

______________________________________________

Why I Want To Accomplish This Goal:

______________________________________________

______________________________________________

______________________________________________

______________________________________________

______________________________________________

By When: ______________________________

What do I have to do to achieve this goal?

______________________________________________

______________________________________________

______________________________________________

______________________________________________

Date Achieved: ______________________

What I Felt When I Accomplished My Goal:

______________________

______________________

______________________

______________________

______________________

______________________

______________________

Paste Your Achievement Photo Here

Bucket List Goal #34: ______________________

Describe The Goal:

______________________________________

______________________________________

______________________________________

______________________________________

______________________________________

Why I Want To Accomplish This Goal:

______________________________________

______________________________________

______________________________________

______________________________________

______________________________________

By When: ______________________________

What do I have to do to achieve this goal?

______________________________________

______________________________________

______________________________________

______________________________________

Date Achieved: ______________________________

What I Felt When I Accomplished My Goal:

______________________________________________

______________________________________________

______________________________________________

______________________________________________

______________________________________________

______________________________________________

______________________________________________

Paste Your Achievement Photo Here

Bucket List Goal #35: ____________________

Describe The Goal:

______________________________________

______________________________________

______________________________________

______________________________________

______________________________________

Why I Want To Accomplish This Goal:

______________________________________

______________________________________

______________________________________

______________________________________

______________________________________

By When: ____________________________

What do I have to do to achieve this goal?

______________________________________

______________________________________

______________________________________

______________________________________

Date Achieved: ______________________________

What I Felt When I Accomplished My Goal:

_____________________________________________

_____________________________________________

_____________________________________________

_____________________________________________

_____________________________________________

_____________________________________________

_____________________________________________

Paste Your Achievement Photo Here

Bucket List Goal #36: ______________________

Describe The Goal:

________________________________________

________________________________________

________________________________________

________________________________________

________________________________________

Why I Want To Accomplish This Goal:

________________________________________

________________________________________

________________________________________

________________________________________

________________________________________

By When: ______________________________

What do I have to do to achieve this goal?

________________________________________

________________________________________

________________________________________

________________________________________

Date Achieved: ______________________________

What I Felt When I Accomplished My Goal:

______________________________

______________________________

______________________________

______________________________

______________________________

______________________________

______________________________

Paste Your Achievement Photo Here

Bucket List Goal #37: ______________________

Describe The Goal:

________________________________________

________________________________________

________________________________________

________________________________________

________________________________________

Why I Want To Accomplish This Goal:

________________________________________

________________________________________

________________________________________

________________________________________

________________________________________

By When: ________________________________

What do I have to do to achieve this goal?

________________________________________

________________________________________

________________________________________

________________________________________

Date Achieved: ______________________________

What I Felt When I Accomplished My Goal:

______________________________________________

______________________________________________

______________________________________________

______________________________________________

______________________________________________

______________________________________________

______________________________________________

Paste Your Achievement Photo Here

Bucket List Goal #38: ______________________

Describe The Goal:

______________________________________

______________________________________

______________________________________

______________________________________

______________________________________

Why I Want To Accomplish This Goal:

______________________________________

______________________________________

______________________________________

______________________________________

______________________________________

By When: ______________________________

What do I have to do to achieve this goal?

______________________________________

______________________________________

______________________________________

______________________________________

Date Achieved: ____________________

What I Felt When I Accomplished My Goal:

____________________

____________________

____________________

____________________

____________________

____________________

____________________

Paste Your Achievement Photo Here

Bucket List Goal #39: ______________________

Describe The Goal:

________________________________________

________________________________________

________________________________________

________________________________________

________________________________________

Why I Want To Accomplish This Goal:

________________________________________

________________________________________

________________________________________

________________________________________

________________________________________

By When: ______________________________

What do I have to do to achieve this goal?

________________________________________

________________________________________

________________________________________

________________________________________

Date Achieved: ____________________

What I Felt When I Accomplished My Goal:

____________________

____________________

____________________

____________________

____________________

____________________

____________________

Paste Your Achievement Photo Here

Bucket List Goal #40: ______________________

Describe The Goal:

______________________________________________

______________________________________________

______________________________________________

______________________________________________

______________________________________________

Why I Want To Accomplish This Goal:

______________________________________________

______________________________________________

______________________________________________

______________________________________________

______________________________________________

By When: ________________________________

What do I have to do to achieve this goal?

______________________________________________

______________________________________________

______________________________________________

______________________________________________

Date Achieved: ____________________

What I Felt When I Accomplished My Goal:

____________________

____________________

____________________

____________________

____________________

____________________

____________________

Paste Your Achievement Photo Here

Bucket List Goal #41: ____________________

Describe The Goal:

____________________

____________________

____________________

____________________

____________________

Why I Want To Accomplish This Goal:

____________________

____________________

____________________

____________________

____________________

By When: ____________________

What do I have to do to achieve this goal?

____________________

____________________

____________________

____________________

Date Achieved: ______________________________

What I Felt When I Accomplished My Goal:

______________________________________________

______________________________________________

______________________________________________

______________________________________________

______________________________________________

______________________________________________

______________________________________________

Paste Your Achievement Photo Here

Bucket List Goal #42: ______________________

Describe The Goal:

______________________________________

______________________________________

______________________________________

______________________________________

______________________________________

Why I Want To Accomplish This Goal:

______________________________________

______________________________________

______________________________________

______________________________________

______________________________________

By When: ______________________________

What do I have to do to achieve this goal?

______________________________________

______________________________________

______________________________________

______________________________________

Date Achieved: ______________________________

What I Felt When I Accomplished My Goal:

______________________________________________

______________________________________________

______________________________________________

______________________________________________

______________________________________________

______________________________________________

______________________________________________

Paste Your Achievement Photo Here

Bucket List Goal #43: ____________________

Describe The Goal:

________________________________________

________________________________________

________________________________________

________________________________________

________________________________________

Why I Want To Accomplish This Goal:

________________________________________

________________________________________

________________________________________

________________________________________

________________________________________

By When: ____________________

What do I have to do to achieve this goal?

________________________________________

________________________________________

________________________________________

________________________________________

Date Achieved: ______________________________

What I Felt When I Accomplished My Goal:

____________________________________________

____________________________________________

____________________________________________

____________________________________________

____________________________________________

____________________________________________

____________________________________________

Paste Your Achievement Photo Here

Bucket List Goal #44: ______________________

Describe The Goal:

______________________________________

______________________________________

______________________________________

______________________________________

______________________________________

Why I Want To Accomplish This Goal:

______________________________________

______________________________________

______________________________________

______________________________________

______________________________________

By When: ______________________________

What do I have to do to achieve this goal?

______________________________________

______________________________________

______________________________________

______________________________________

Date Achieved: ________________________

What I Felt When I Accomplished My Goal:

________________________

________________________

________________________

________________________

________________________

________________________

________________________

Paste Your Achievement Photo Here

Bucket List Goal #45: ____________________

Describe The Goal:

______________________________________

______________________________________

______________________________________

______________________________________

______________________________________

Why I Want To Accomplish This Goal:

______________________________________

______________________________________

______________________________________

______________________________________

______________________________________

By When: ____________________________

What do I have to do to achieve this goal?

______________________________________

______________________________________

______________________________________

______________________________________

Date Achieved: ______________________________

What I Felt When I Accomplished My Goal:

___________________________________________

___________________________________________

___________________________________________

___________________________________________

___________________________________________

___________________________________________

___________________________________________

Paste Your Achievement Photo Here

Bucket List Goal #46: ______________________

Describe The Goal:

______________________________________

______________________________________

______________________________________

______________________________________

______________________________________

Why I Want To Accomplish This Goal:

______________________________________

______________________________________

______________________________________

______________________________________

______________________________________

By When: ______________________________

What do I have to do to achieve this goal?

______________________________________

______________________________________

______________________________________

______________________________________

Date Achieved: ______________________________

What I Felt When I Accomplished My Goal:

______________________________________________

______________________________________________

______________________________________________

______________________________________________

______________________________________________

______________________________________________

______________________________________________

Paste Your Achievement Photo Here

Bucket List Goal #47: ______________________

Describe The Goal:

________________________________________

________________________________________

________________________________________

________________________________________

________________________________________

Why I Want To Accomplish This Goal:

________________________________________

________________________________________

________________________________________

________________________________________

________________________________________

By When: ____________________________

What do I have to do to achieve this goal?

________________________________________

________________________________________

________________________________________

________________________________________

Date Achieved: ______________________________

What I Felt When I Accomplished My Goal:

_____________________________________________

_____________________________________________

_____________________________________________

_____________________________________________

_____________________________________________

_____________________________________________

_____________________________________________

Paste Your Achievement Photo Here

Bucket List Goal #48: ______________________

Describe The Goal:

________________________________________

________________________________________

________________________________________

________________________________________

________________________________________

Why I Want To Accomplish This Goal:

________________________________________

________________________________________

________________________________________

________________________________________

________________________________________

By When: ______________________________

What do I have to do to achieve this goal?

________________________________________

________________________________________

________________________________________

________________________________________

Date Achieved: ______________________________

What I Felt When I Accomplished My Goal:

______________________________________________

______________________________________________

______________________________________________

______________________________________________

______________________________________________

______________________________________________

______________________________________________

Paste Your Achievement Photo Here

Bucket List Goal #49: ______________________

Describe The Goal:

______________________________________

______________________________________

______________________________________

______________________________________

______________________________________

Why I Want To Accomplish This Goal:

______________________________________

______________________________________

______________________________________

______________________________________

______________________________________

By When: ______________________________

What do I have to do to achieve this goal?

______________________________________

______________________________________

______________________________________

______________________________________

Date Achieved: ______________________________

What I Felt When I Accomplished My Goal:

____________________________________________

____________________________________________

____________________________________________

____________________________________________

____________________________________________

____________________________________________

____________________________________________

Paste Your Achievement Photo Here

Bucket List Goal #50: ______________________

Describe The Goal:

______________________________________

______________________________________

______________________________________

______________________________________

______________________________________

Why I Want To Accomplish This Goal:

______________________________________

______________________________________

______________________________________

______________________________________

______________________________________

By When: ______________________________

What do I have to do to achieve this goal?

______________________________________

______________________________________

______________________________________

______________________________________

Date Achieved: ______________________________

What I Felt When I Accomplished My Goal:

______________________________________________

______________________________________________

______________________________________________

______________________________________________

______________________________________________

______________________________________________

______________________________________________

Paste Your Achievement Photo Here

Bucket List Goal #51: ______________________

Describe The Goal:

______________________________________

______________________________________

______________________________________

______________________________________

______________________________________

Why I Want To Accomplish This Goal:

______________________________________

______________________________________

______________________________________

______________________________________

______________________________________

By When: ______________________________

What do I have to do to achieve this goal?

______________________________________

______________________________________

______________________________________

______________________________________

Date Achieved: ______________________________

What I Felt When I Accomplished My Goal:

______________________________________________

______________________________________________

______________________________________________

______________________________________________

______________________________________________

______________________________________________

______________________________________________

Paste Your Achievement Photo Here

Bucket List Goal #52: ______________________

Describe The Goal:

______________________________________

______________________________________

______________________________________

______________________________________

______________________________________

Why I Want To Accomplish This Goal:

______________________________________

______________________________________

______________________________________

______________________________________

______________________________________

By When: ____________________________

What do I have to do to achieve this goal?

______________________________________

______________________________________

______________________________________

______________________________________

Date Achieved: ______________________________

What I Felt When I Accomplished My Goal:

______________________________________________

______________________________________________

______________________________________________

______________________________________________

______________________________________________

______________________________________________

______________________________________________

Paste Your Achievement Photo Here

Bucket List Goal #53: ______________________

Describe The Goal:

______________________________________

______________________________________

______________________________________

______________________________________

______________________________________

Why I Want To Accomplish This Goal:

______________________________________

______________________________________

______________________________________

______________________________________

______________________________________

By When: ______________________________

What do I have to do to achieve this goal?

______________________________________

______________________________________

______________________________________

______________________________________

Date Achieved: ______________________________

What I Felt When I Accomplished My Goal:

______________________________________________

______________________________________________

______________________________________________

______________________________________________

______________________________________________

______________________________________________

______________________________________________

Paste Your Achievement Photo Here

Bucket List Goal #54: ______________________

Describe The Goal:

________________________________________

________________________________________

________________________________________

________________________________________

________________________________________

Why I Want To Accomplish This Goal:

________________________________________

________________________________________

________________________________________

________________________________________

________________________________________

By When: ______________________________

What do I have to do to achieve this goal?

________________________________________

________________________________________

________________________________________

________________________________________

Date Achieved: ______________________________

What I Felt When I Accomplished My Goal:

______________________________________________

______________________________________________

______________________________________________

______________________________________________

______________________________________________

______________________________________________

______________________________________________

Paste Your Achievement Photo Here

Bucket List Goal #55: ____________________

Describe The Goal:

______________________________

______________________________

______________________________

______________________________

______________________________

Why I Want To Accomplish This Goal:

______________________________

______________________________

______________________________

______________________________

______________________________

By When: ____________________

What do I have to do to achieve this goal?

______________________________

______________________________

______________________________

______________________________

Date Achieved: ______________________________

What I Felt When I Accomplished My Goal:

______________________________________________

______________________________________________

______________________________________________

______________________________________________

______________________________________________

______________________________________________

______________________________________________

Paste Your Achievement Photo Here

Bucket List Goal #56: ______________________

Describe The Goal:

______________________________________________

______________________________________________

______________________________________________

______________________________________________

______________________________________________

Why I Want To Accomplish This Goal:

______________________________________________

______________________________________________

______________________________________________

______________________________________________

______________________________________________

By When: ______________________________

What do I have to do to achieve this goal?

______________________________________________

______________________________________________

______________________________________________

______________________________________________

Date Achieved: ______________________________

What I Felt When I Accomplished My Goal:

______________________________

______________________________

______________________________

______________________________

______________________________

______________________________

______________________________

Paste Your Achievement Photo Here

Bucket List Goal #57: ______________________

Describe The Goal:

______________________________________

______________________________________

______________________________________

______________________________________

______________________________________

Why I Want To Accomplish This Goal:

______________________________________

______________________________________

______________________________________

______________________________________

______________________________________

By When: ______________________________

What do I have to do to achieve this goal?

______________________________________

______________________________________

______________________________________

______________________________________

Date Achieved: ______________________________

What I Felt When I Accomplished My Goal:

______________________________________________

______________________________________________

______________________________________________

______________________________________________

______________________________________________

______________________________________________

______________________________________________

Paste Your Achievement Photo Here

Bucket List Goal #58: ____________________

Describe The Goal:

____________________

____________________

____________________

____________________

____________________

Why I Want To Accomplish This Goal:

____________________

____________________

____________________

____________________

____________________

By When: ____________________

What do I have to do to achieve this goal?

____________________

____________________

____________________

____________________

Date Achieved: ______________________________

What I Felt When I Accomplished My Goal:

______________________________________________

______________________________________________

______________________________________________

______________________________________________

______________________________________________

______________________________________________

______________________________________________

Paste Your Achievement Photo Here

Bucket List Goal #59: ____________________

Describe The Goal:

______________________________

______________________________

______________________________

______________________________

______________________________

Why I Want To Accomplish This Goal:

______________________________

______________________________

______________________________

______________________________

______________________________

By When: ____________________

What do I have to do to achieve this goal?

______________________________

______________________________

______________________________

______________________________

Date Achieved: ____________________________

What I Felt When I Accomplished My Goal:

________________________________________

________________________________________

________________________________________

________________________________________

________________________________________

________________________________________

________________________________________

Paste Your Achievement Photo Here

Bucket List Goal #60: ______________________

Describe The Goal:

______________________________________

______________________________________

______________________________________

______________________________________

______________________________________

Why I Want To Accomplish This Goal:

______________________________________

______________________________________

______________________________________

______________________________________

______________________________________

By When: ______________________

What do I have to do to achieve this goal?

______________________________________

______________________________________

______________________________________

______________________________________

Date Achieved: ______________________________

What I Felt When I Accomplished My Goal:

_____________________________________________

_____________________________________________

_____________________________________________

_____________________________________________

_____________________________________________

_____________________________________________

_____________________________________________

Paste Your Achievement Photo Here

Bucket List Goal #61: ______________________

Describe The Goal:

________________________________________

________________________________________

________________________________________

________________________________________

________________________________________

Why I Want To Accomplish This Goal:

________________________________________

________________________________________

________________________________________

________________________________________

________________________________________

By When: ______________________________

What do I have to do to achieve this goal?

________________________________________

________________________________________

________________________________________

________________________________________

Date Achieved: ______________________________

What I Felt When I Accomplished My Goal:

______________________________

______________________________

______________________________

______________________________

______________________________

______________________________

______________________________

Paste Your Achievement Photo Here

Bucket List Goal #62: ______________________

Describe The Goal:

________________________________________

________________________________________

________________________________________

________________________________________

________________________________________

Why I Want To Accomplish This Goal:

________________________________________

________________________________________

________________________________________

________________________________________

________________________________________

By When: ______________________________

What do I have to do to achieve this goal?

________________________________________

________________________________________

________________________________________

________________________________________

Date Achieved: ______________________________

What I Felt When I Accomplished My Goal:

______________________________________________

______________________________________________

______________________________________________

______________________________________________

______________________________________________

______________________________________________

______________________________________________

Paste Your Achievement Photo Here

Bucket List Goal #63: ______________________

Describe The Goal:

________________________________________

________________________________________

________________________________________

________________________________________

________________________________________

Why I Want To Accomplish This Goal:

________________________________________

________________________________________

________________________________________

________________________________________

________________________________________

By When: ______________________________

What do I have to do to achieve this goal?

________________________________________

________________________________________

________________________________________

________________________________________

Date Achieved: ______________________________

What I Felt When I Accomplished My Goal:

______________________________________________

______________________________________________

______________________________________________

______________________________________________

______________________________________________

______________________________________________

______________________________________________

Paste Your Achievement Photo Here

Bucket List Goal #64: ____________________

Describe The Goal:

____________________

____________________

____________________

____________________

____________________

Why I Want To Accomplish This Goal:

____________________

____________________

____________________

____________________

____________________

By When: ____________________

What do I have to do to achieve this goal?

____________________

____________________

____________________

____________________

Date Achieved: ______________________________

What I Felt When I Accomplished My Goal:

______________________________

______________________________

______________________________

______________________________

______________________________

______________________________

______________________________

Paste Your Achievement Photo Here

Bucket List Goal #65: ______________________

Describe The Goal:

______________________________________

______________________________________

______________________________________

______________________________________

______________________________________

Why I Want To Accomplish This Goal:

______________________________________

______________________________________

______________________________________

______________________________________

______________________________________

By When: ______________________________

What do I have to do to achieve this goal?

______________________________________

______________________________________

______________________________________

______________________________________

Date Achieved: ______________________________

What I Felt When I Accomplished My Goal:

______________________________________________

______________________________________________

______________________________________________

______________________________________________

______________________________________________

______________________________________________

______________________________________________

Paste Your Achievement Photo Here

Bucket List Goal #66: ____________________

Describe The Goal:

____________________

____________________

____________________

____________________

____________________

Why I Want To Accomplish This Goal:

____________________

____________________

____________________

____________________

____________________

By When: ____________________

What do I have to do to achieve this goal?

____________________

____________________

____________________

____________________

Date Achieved: ______________________________

What I Felt When I Accomplished My Goal:

______________________________________________

______________________________________________

______________________________________________

______________________________________________

______________________________________________

______________________________________________

______________________________________________

Paste Your Achievement Photo Here

Bucket List Goal #67: ____________________

Describe The Goal:

______________________________

______________________________

______________________________

______________________________

______________________________

Why I Want To Accomplish This Goal:

______________________________

______________________________

______________________________

______________________________

______________________________

By When: ____________________

What do I have to do to achieve this goal?

______________________________

______________________________

______________________________

______________________________

Date Achieved: ______________________________

What I Felt When I Accomplished My Goal:

______________________________

______________________________

______________________________

______________________________

______________________________

______________________________

______________________________

Paste Your Achievement Photo Here

Bucket List Goal #68: ______________________

Describe The Goal:

______________________________________

______________________________________

______________________________________

______________________________________

______________________________________

Why I Want To Accomplish This Goal:

______________________________________

______________________________________

______________________________________

______________________________________

______________________________________

By When: ______________________________

What do I have to do to achieve this goal?

______________________________________

______________________________________

______________________________________

______________________________________

Date Achieved: ______________________________

What I Felt When I Accomplished My Goal:

______________________________________________

______________________________________________

______________________________________________

______________________________________________

______________________________________________

______________________________________________

______________________________________________

Paste Your Achievement Photo Here

Bucket List Goal #69: ______________________

Describe The Goal:

________________________________________

________________________________________

________________________________________

________________________________________

________________________________________

Why I Want To Accomplish This Goal:

________________________________________

________________________________________

________________________________________

________________________________________

________________________________________

By When: ______________________________

What do I have to do to achieve this goal?

________________________________________

________________________________________

________________________________________

________________________________________

Date Achieved: ___________________________

What I Felt When I Accomplished My Goal:

___________________________

___________________________

___________________________

___________________________

___________________________

___________________________

___________________________

Paste Your Achievement Photo Here

Bucket List Goal #70: ______________________

Describe The Goal:

______________________________________

______________________________________

______________________________________

______________________________________

______________________________________

Why I Want To Accomplish This Goal:

______________________________________

______________________________________

______________________________________

______________________________________

______________________________________

By When: ______________________________

What do I have to do to achieve this goal?

______________________________________

______________________________________

______________________________________

______________________________________

Date Achieved: ______________________________

What I Felt When I Accomplished My Goal:

______________________________________________

______________________________________________

______________________________________________

______________________________________________

______________________________________________

______________________________________________

______________________________________________

Paste Your Achievement Photo Here

Bucket List Goal #71: ____________________

Describe The Goal:

____________________

____________________

____________________

____________________

____________________

Why I Want To Accomplish This Goal:

____________________

____________________

____________________

____________________

____________________

By When: ____________________

What do I have to do to achieve this goal?

____________________

____________________

____________________

____________________

Date Achieved: ______________________

What I Felt When I Accomplished My Goal:

______________________

______________________

______________________

______________________

______________________

______________________

______________________

Paste Your Achievement Photo Here

Bucket List Goal #72: ______________________

Describe The Goal:

________________________________________

________________________________________

________________________________________

________________________________________

________________________________________

Why I Want To Accomplish This Goal:

________________________________________

________________________________________

________________________________________

________________________________________

________________________________________

By When: ______________________________

What do I have to do to achieve this goal?

________________________________________

________________________________________

________________________________________

________________________________________

Date Achieved: ______________________________

What I Felt When I Accomplished My Goal:

______________________________________________

______________________________________________

______________________________________________

______________________________________________

______________________________________________

______________________________________________

______________________________________________

Paste Your Achievement Photo Here

Bucket List Goal #73: ______________________

Describe The Goal:

______________________________________

______________________________________

______________________________________

______________________________________

______________________________________

Why I Want To Accomplish This Goal:

______________________________________

______________________________________

______________________________________

______________________________________

______________________________________

By When: ______________________

What do I have to do to achieve this goal?

______________________________________

______________________________________

______________________________________

______________________________________

Date Achieved: ______________________________

What I Felt When I Accomplished My Goal:

______________________________________________

______________________________________________

______________________________________________

______________________________________________

______________________________________________

______________________________________________

______________________________________________

Paste Your Achievement Photo Here

Bucket List Goal #74: ______________________

Describe The Goal:

______________________________________________

______________________________________________

______________________________________________

______________________________________________

______________________________________________

Why I Want To Accomplish This Goal:

______________________________________________

______________________________________________

______________________________________________

______________________________________________

______________________________________________

By When: ______________________________

What do I have to do to achieve this goal?

______________________________________________

______________________________________________

______________________________________________

______________________________________________

Date Achieved: ______________________________

What I Felt When I Accomplished My Goal:

______________________________________________

______________________________________________

______________________________________________

______________________________________________

______________________________________________

______________________________________________

______________________________________________

Paste Your Achievement Photo Here

Bucket List Goal #75: ______________________

Describe The Goal:

________________________________________

________________________________________

________________________________________

________________________________________

________________________________________

Why I Want To Accomplish This Goal:

________________________________________

________________________________________

________________________________________

________________________________________

________________________________________

By When: ______________________________

What do I have to do to achieve this goal?

________________________________________

________________________________________

________________________________________

________________________________________

Date Achieved: ______________________________

What I Felt When I Accomplished My Goal:

______________________________

______________________________

______________________________

______________________________

______________________________

______________________________

______________________________

Paste Your Achievement Photo Here

Bucket List Goal #76: ______________________

Describe The Goal:

________________________________________

________________________________________

________________________________________

________________________________________

________________________________________

Why I Want To Accomplish This Goal:

________________________________________

________________________________________

________________________________________

________________________________________

________________________________________

By When: ______________________________

What do I have to do to achieve this goal?

________________________________________

________________________________________

________________________________________

________________________________________

Date Achieved: ______________________________

What I Felt When I Accomplished My Goal:

_____________________________________________

_____________________________________________

_____________________________________________

_____________________________________________

_____________________________________________

_____________________________________________

_____________________________________________

Paste Your Achievement Photo Here

Bucket List Goal #77: ______________________

Describe The Goal:

______________________________________

______________________________________

______________________________________

______________________________________

______________________________________

Why I Want To Accomplish This Goal:

______________________________________

______________________________________

______________________________________

______________________________________

______________________________________

By When: ______________________________

What do I have to do to achieve this goal?

______________________________________

______________________________________

______________________________________

______________________________________

Date Achieved: ______________________________

What I Felt When I Accomplished My Goal:

______________________________________________

______________________________________________

______________________________________________

______________________________________________

______________________________________________

______________________________________________

______________________________________________

Paste Your Achievement Photo Here

Bucket List Goal #78: ______________________

Describe The Goal:

________________________________________

________________________________________

________________________________________

________________________________________

________________________________________

Why I Want To Accomplish This Goal:

________________________________________

________________________________________

________________________________________

________________________________________

________________________________________

By When: ______________________________

What do I have to do to achieve this goal?

________________________________________

________________________________________

________________________________________

________________________________________

Date Achieved: ______________________________

What I Felt When I Accomplished My Goal:

______________________________________________

______________________________________________

______________________________________________

______________________________________________

______________________________________________

______________________________________________

______________________________________________

Paste Your Achievement Photo Here

Bucket List Goal #79: ______________________

Describe The Goal:

______________________________________

______________________________________

______________________________________

______________________________________

______________________________________

Why I Want To Accomplish This Goal:

______________________________________

______________________________________

______________________________________

______________________________________

______________________________________

By When: ______________________________

What do I have to do to achieve this goal?

______________________________________

______________________________________

______________________________________

______________________________________

Date Achieved: ______________________________

What I Felt When I Accomplished My Goal:

______________________________________________

______________________________________________

______________________________________________

______________________________________________

______________________________________________

______________________________________________

______________________________________________

Paste Your Achievement Photo Here

Bucket List Goal #80: ______________________

Describe The Goal:

______________________________________

______________________________________

______________________________________

______________________________________

______________________________________

Why I Want To Accomplish This Goal:

______________________________________

______________________________________

______________________________________

______________________________________

______________________________________

By When: ______________________________

What do I have to do to achieve this goal?

______________________________________

______________________________________

______________________________________

______________________________________

Date Achieved: ______________________________

What I Felt When I Accomplished My Goal:

____________________________________________

____________________________________________

____________________________________________

____________________________________________

____________________________________________

____________________________________________

____________________________________________

Paste Your Achievement Photo Here

Bucket List Goal #81: ______________________

Describe The Goal:

______________________________________

______________________________________

______________________________________

______________________________________

______________________________________

Why I Want To Accomplish This Goal:

______________________________________

______________________________________

______________________________________

______________________________________

______________________________________

By When: ______________________

What do I have to do to achieve this goal?

______________________________________

______________________________________

______________________________________

______________________________________

Date Achieved: ______________________________

What I Felt When I Accomplished My Goal:

______________________________________________

______________________________________________

______________________________________________

______________________________________________

______________________________________________

______________________________________________

______________________________________________

Paste Your Achievement Photo Here

Bucket List Goal #82: ______________________

Describe The Goal:

______________________________________

______________________________________

______________________________________

______________________________________

______________________________________

Why I Want To Accomplish This Goal:

______________________________________

______________________________________

______________________________________

______________________________________

______________________________________

By When: ______________________________

What do I have to do to achieve this goal?

______________________________________

______________________________________

______________________________________

______________________________________

Date Achieved: ______________________________

What I Felt When I Accomplished My Goal:

______________________________________________

______________________________________________

______________________________________________

______________________________________________

______________________________________________

______________________________________________

______________________________________________

Paste Your Achievement Photo Here

Bucket List Goal #83: ______________________

Describe The Goal:

________________________________________

________________________________________

________________________________________

________________________________________

________________________________________

Why I Want To Accomplish This Goal:

________________________________________

________________________________________

________________________________________

________________________________________

________________________________________

By When: ______________________________

What do I have to do to achieve this goal?

________________________________________

________________________________________

________________________________________

________________________________________

Date Achieved: ______________________________

What I Felt When I Accomplished My Goal:

______________________________

______________________________

______________________________

______________________________

______________________________

______________________________

______________________________

Paste Your Achievement Photo Here

Bucket List Goal #84: ____________________

Describe The Goal:

______________________________

______________________________

______________________________

______________________________

______________________________

Why I Want To Accomplish This Goal:

______________________________

______________________________

______________________________

______________________________

______________________________

By When: ____________________

What do I have to do to achieve this goal?

______________________________

______________________________

______________________________

______________________________

Date Achieved: ____________________

What I Felt When I Accomplished My Goal:

____________________

____________________

____________________

____________________

____________________

____________________

____________________

Paste Your Achievement Photo Here

Bucket List Goal #85: ____________________

Describe The Goal:

____________________

____________________

____________________

____________________

____________________

Why I Want To Accomplish This Goal:

____________________

____________________

____________________

____________________

____________________

By When: ____________________

What do I have to do to achieve this goal?

____________________

____________________

____________________

____________________

Date Achieved: ______________________________

What I Felt When I Accomplished My Goal:

______________________________

______________________________

______________________________

______________________________

______________________________

______________________________

______________________________

Paste Your Achievement Photo Here

Bucket List Goal #86: ____________________

Describe The Goal:

________________________________________

________________________________________

________________________________________

________________________________________

________________________________________

Why I Want To Accomplish This Goal:

________________________________________

________________________________________

________________________________________

________________________________________

________________________________________

By When: ____________________________

What do I have to do to achieve this goal?

________________________________________

________________________________________

________________________________________

________________________________________

Date Achieved: ___________________________

What I Felt When I Accomplished My Goal:

___________________________________________

___________________________________________

___________________________________________

___________________________________________

___________________________________________

___________________________________________

___________________________________________

Paste Your Achievement Photo Here

Bucket List Goal #87: ______________________________

Describe The Goal:

______________________________________________

______________________________________________

______________________________________________

______________________________________________

______________________________________________

Why I Want To Accomplish This Goal:

______________________________________________

______________________________________________

______________________________________________

______________________________________________

______________________________________________

By When: ______________________________________

What do I have to do to achieve this goal?

______________________________________________

______________________________________________

______________________________________________

______________________________________________

Date Achieved: ______________________________

What I Felt When I Accomplished My Goal:

______________________________________________

______________________________________________

______________________________________________

______________________________________________

______________________________________________

______________________________________________

______________________________________________

Paste Your Achievement Photo Here

Bucket List Goal #88: ______________________

Describe The Goal:

________________________________________

________________________________________

________________________________________

________________________________________

________________________________________

Why I Want To Accomplish This Goal:

________________________________________

________________________________________

________________________________________

________________________________________

________________________________________

By When: ______________________________

What do I have to do to achieve this goal?

________________________________________

________________________________________

________________________________________

________________________________________

Date Achieved: ______________________________

What I Felt When I Accomplished My Goal:

______________________________________________

______________________________________________

______________________________________________

______________________________________________

______________________________________________

______________________________________________

______________________________________________

Paste Your Achievement Photo Here

Bucket List Goal #89: ____________________

Describe The Goal:

________________________________________

________________________________________

________________________________________

________________________________________

________________________________________

Why I Want To Accomplish This Goal:

________________________________________

________________________________________

________________________________________

________________________________________

________________________________________

By When: ____________________

What do I have to do to achieve this goal?

________________________________________

________________________________________

________________________________________

________________________________________

Date Achieved: ______________________________

What I Felt When I Accomplished My Goal:

______________________________

______________________________

______________________________

______________________________

______________________________

______________________________

______________________________

Paste Your Achievement Photo Here

Bucket List Goal #90: ______________________

Describe The Goal:

______________________________________________

______________________________________________

______________________________________________

______________________________________________

______________________________________________

Why I Want To Accomplish This Goal:

______________________________________________

______________________________________________

______________________________________________

______________________________________________

______________________________________________

By When: ______________________________

What do I have to do to achieve this goal?

______________________________________________

______________________________________________

______________________________________________

______________________________________________

Date Achieved: ______________________________

What I Felt When I Accomplished My Goal:

______________________________________________

______________________________________________

______________________________________________

______________________________________________

______________________________________________

______________________________________________

______________________________________________

Paste Your Achievement Photo Here

Bucket List Goal #91: ______________________

Describe The Goal:

______________________________________

______________________________________

______________________________________

______________________________________

______________________________________

Why I Want To Accomplish This Goal:

______________________________________

______________________________________

______________________________________

______________________________________

______________________________________

By When: ______________________________

What do I have to do to achieve this goal?

______________________________________

______________________________________

______________________________________

______________________________________

Date Achieved: ______________________

What I Felt When I Accomplished My Goal:

______________________

______________________

______________________

______________________

______________________

______________________

______________________

Paste Your Achievement Photo Here

Bucket List Goal #92: ____________________

Describe The Goal:

____________________

____________________

____________________

____________________

____________________

Why I Want To Accomplish This Goal:

____________________

____________________

____________________

____________________

____________________

By When: ____________________

What do I have to do to achieve this goal?

____________________

____________________

____________________

____________________

Date Achieved: ______________________________

What I Felt When I Accomplished My Goal:

______________________________________________

______________________________________________

______________________________________________

______________________________________________

______________________________________________

______________________________________________

______________________________________________

Paste Your Achievement Photo Here

Bucket List Goal #93: ______________________

Describe The Goal:

______________________________________

______________________________________

______________________________________

______________________________________

______________________________________

Why I Want To Accomplish This Goal:

______________________________________

______________________________________

______________________________________

______________________________________

______________________________________

By When: ______________________________

What do I have to do to achieve this goal?

______________________________________

______________________________________

______________________________________

______________________________________

Date Achieved: ______________________________

What I Felt When I Accomplished My Goal:

______________________________

______________________________

______________________________

______________________________

______________________________

______________________________

______________________________

Paste Your Achievement Photo Here

Bucket List Goal #94: ______________________

Describe The Goal:

______________________________________

______________________________________

______________________________________

______________________________________

______________________________________

Why I Want To Accomplish This Goal:

______________________________________

______________________________________

______________________________________

______________________________________

______________________________________

By When: ______________________________

What do I have to do to achieve this goal?

______________________________________

______________________________________

______________________________________

______________________________________

Date Achieved: ______________________________

What I Felt When I Accomplished My Goal:

______________________________________________

______________________________________________

______________________________________________

______________________________________________

______________________________________________

______________________________________________

______________________________________________

Paste Your Achievement Photo Here

Bucket List Goal #95: ______________________

Describe The Goal:

______________________________________________

______________________________________________

______________________________________________

______________________________________________

______________________________________________

Why I Want To Accomplish This Goal:

______________________________________________

______________________________________________

______________________________________________

______________________________________________

______________________________________________

By When: ______________________________

What do I have to do to achieve this goal?

______________________________________________

______________________________________________

______________________________________________

______________________________________________

Date Achieved: ______________________________

What I Felt When I Accomplished My Goal:

______________________________

______________________________

______________________________

______________________________

______________________________

______________________________

______________________________

Paste Your Achievement Photo Here

Bucket List Goal #96: ______________________

Describe The Goal:

______________________________________

______________________________________

______________________________________

______________________________________

______________________________________

Why I Want To Accomplish This Goal:

______________________________________

______________________________________

______________________________________

______________________________________

______________________________________

By When: ______________________________

What do I have to do to achieve this goal?

______________________________________

______________________________________

______________________________________

______________________________________

Date Achieved: ______________________________

What I Felt When I Accomplished My Goal:

______________________________

______________________________

______________________________

______________________________

______________________________

______________________________

______________________________

Paste Your Achievement Photo Here

Bucket List Goal #97: ______________________

Describe The Goal:

______________________

______________________

______________________

______________________

______________________

Why I Want To Accomplish This Goal:

______________________

______________________

______________________

______________________

______________________

By When: ______________________

What do I have to do to achieve this goal?

______________________

______________________

______________________

______________________

Date Achieved: ______________________________

What I Felt When I Accomplished My Goal:

______________________________

______________________________

______________________________

______________________________

______________________________

______________________________

______________________________

Paste Your Achievement Photo Here

Bucket List Goal #98: ______________________

Describe The Goal:

______________________________________

______________________________________

______________________________________

______________________________________

______________________________________

Why I Want To Accomplish This Goal:

______________________________________

______________________________________

______________________________________

______________________________________

______________________________________

By When: ______________________________

What do I have to do to achieve this goal?

______________________________________

______________________________________

______________________________________

______________________________________

Date Achieved: ________________________________

What I Felt When I Accomplished My Goal:

______________________________________________

______________________________________________

______________________________________________

______________________________________________

______________________________________________

______________________________________________

______________________________________________

Paste Your Achievement Photo Here

Bucket List Goal #99: ______________________

Describe The Goal:

______________________________________

______________________________________

______________________________________

______________________________________

______________________________________

Why I Want To Accomplish This Goal:

______________________________________

______________________________________

______________________________________

______________________________________

______________________________________

By When: ______________________________

What do I have to do to achieve this goal?

______________________________________

______________________________________

______________________________________

______________________________________

Date Achieved: ______________________________

What I Felt When I Accomplished My Goal:

______________________________________________

______________________________________________

______________________________________________

______________________________________________

______________________________________________

______________________________________________

______________________________________________

Paste Your Achievement Photo Here

Bucket List Goal #100: ______________________

Describe The Goal:

____________________________________________

____________________________________________

____________________________________________

____________________________________________

____________________________________________

Why I Want To Accomplish This Goal:

____________________________________________

____________________________________________

____________________________________________

____________________________________________

____________________________________________

By When: __________________________________

What do I have to do to achieve this goal?

____________________________________________

____________________________________________

____________________________________________

____________________________________________

Date Achieved: ______________________________

What I Felt When I Accomplished My Goal:

______________________________

______________________________

______________________________

______________________________

______________________________

______________________________

______________________________

Paste Your Achievement Photo Here

Made in the USA
Lexington, KY
13 December 2018